It's How Old?

Harcourt
SCHOOL PUBLISHERS

Orlando   Austin   New York   San Diego   Toronto   London

Visit *The Learning Site!*
www.harcourtschool.com

Introduction

If you were a detective at a crime scene, you would need to gather evidence that would help you solve the crime. You would also need to ask yourself several questions about the crime. When did the crime happen? Who was there when the crime occurred? How did the crime take place?

If you were a *paleontologist* (fossil scientist), you would ask the same kinds of questions about a fossil you found. When did the organism live? What other organisms lived at the same time? How did the organism behave or act when it was alive? Like a detective, you'd be finding clues—clues to what life was like long ago.

A paleontologist might study fossils like the skull shown on the front cover of this book. It's a deinonychus (dy•NAH•nih•kuhs) skull. By looking at the skull, you can tell that this dinosaur was a meat-eater. The teeth are sharply pointed and curve backward, making them better for grabbing and tearing. A deinonychus's teeth were *serrated,* which means that the side of each tooth was like a saw.

The deinonychus was an intelligent meat-eater whose skull had room for a larger brain than most other dinosaur skulls had. In fact, one of the deadly dinosaurs in *Jurassic Park* was actually modeled after the deinonychus, although in the movie it was called a velociraptor (veh•LAH•sih•rap•ter). The velociraptor looked like a deinonychus, but was smaller.

As a paleontologist, you would be curious to know the age of the deinonychus's skull fossil. Because fossils are mostly found in rocks, you would want to learn more about the rocks where the fossil was found.

Earth's Surface Changes

The ground can change very quickly. One day in 1943, near Mexico City, Mexico, a farmer watched in amazement as his land cracked open. Smoke, magma and ash hissed out. A new volcano was starting to grow and was named Paricutín. By the end of the day, Paricutín was 7.5 m (25 ft) high, and it eventually grew to a height of 410 m (1,345 ft).

Other changes in Earth can take a much longer time. The Grand Canyon was carved out very slowly by the Colorado River. It took over six million years to form this magnificent canyon, which is from 0.2 to 29 km (0.1 to 18 mi) wide and, in some places, 1,850 m (6,000 ft) deep.

In 1943, the volcano Paricutín suddenly grew out of a cornfield.

Changes in Earth, whether quick or slow, always follow certain *laws* (statements about how nature works). The forces that change Earth's surface, such as volcanoes and erosion, are the same now as they were a billion years ago.

Because scientists understand how Earth has changed, they can figure out when it changed in certain ways. The deinonychus skull was found in a layer of rock. If scientists can determine how old that layer of rock is, they'll know how old the fossil is.

A Rock Above the Rest

When Paricutín erupted, its ash fell on top of, not under, the cornfield. If the farmer later dug through the ash and found a can, he'd guess that the can was on the ground before the ash landed on the ground.

In the same way, when sediment drifts to the sea floor, it lands on top of, not under, the rock layer at the bottom. New rock forms on top of old rock.

In science, this is called the *law of superposition. Super* means "over" or "above," so superposition means that new rock will take a position on top of existing rock. The law of superposition is the key to learning the age of a fossil.

This is the Grand Canyon. The rocks and the fossils here are newer in the higher layers and older in the lower layers.

Geologists use the law of superposition to find out in what order rock layers were formed. Unfortunately, they can't tell exactly when rock layers formed. A geologist uses *relative dating* to determine which rocks are older than others. *Relative* means "compared to each other."

The Grand Canyon has many layers of sedimentary rock. Relative to the other layers, the topmost layer is the newest.

Old, Older, Oldest

Fossils are found in layers of sedimentary rock. Fossils formed from ancient plants and animals at the same time that sediment was becoming rock. Just like sedimentary rock, fossils found in higher layers are newer than fossils found in lower layers.

The Grand Canyon contains a rock layer called Redwall Limestone. The Redwall Limestone layer has fossils of sea animals, such as clams and snails. The layer above the Redwall Limestone layer is the Supai Formation. Fossils of reptiles and amphibians lie in the shale of the Supai Formation. Using the law of superposition, you can say that the Supai reptiles and amphibians lived later than the Redwall sea animals.

If sedimentary rock has fossils in it, those fossils are the same age as the rock. Therefore, if you can learn the age of the rock, you'll know the age of the fossil.

Laid Down Flat

When you put a slice of bread on a plate, you usually lay it down flat rather than standing it up on one edge. When snow falls, it drifts down and smoothly covers the land. In the same way, when sand settles on the bottom of a lake or sea, it forms a flat layer.

If that sand becomes sedimentary rock, it doesn't rise up into the air the way a volcano does. It lies flat across the ground. Geologists call this the *principle of original horizontality. Original* means "how it was when it formed." Although the rock may start out as a flat layer across the ground, it may not stay that way.

Some sedimentary rocks end up steeply tilted. In Red Rocks Park near Denver, Colorado, huge slabs of red sandstone lean against a mountainside.

The sandstone slabs were raised up by movement of Earth's crust. This area, now in the foothills of the Rocky Mountains, used to be a soggy plain 140 million years ago. A quiet river dropped sand and mud, which over time became sandstone and mudstone. As the Rocky Mountains were pushed upward, they lifted up one side of the rock layer, tilting today's Red Rocks.

When scientists see tilted sedimentary layers such as Red Rocks, they know that a great force pushed one side higher than the other. They also know that the layer was tilted after, not before, it became a rock. The relative age of the sedimentary rock is older than the force that moved it.

At Red Rocks Park, sedimentary layers tilted when the Rocky Mountains were pushed upward.

New Magma in Old Rocks

We most often think of magma as spurting out of the top of a volcano, as it did from Paricutín. However, magma moves below ground much more than it does on the surface. It squeezes through cracks and pushes through tunnels. It squirts out the side of a volcano and splashes out the top.

Travelers can see the Grand Canyon's bottom layer down near the Colorado River. That layer is made of schist (SHIST), which is a type of metamorphic rock. Within the schist lies a great block of granite, which is igneous rock. In the long-ago past, the granite cooled when the magma pushed into the schist.

The *principle of cross-cutting relationships* states that when you find igneous rock pushed inside a layer of other rock, the other rock is older. Read on to see just how helpful this is in learning the ages of rocks and fossils.

Fossil Succession

You've learned that different fossils formed at different times. For example, fossils of the apatosaurus come from a different time than those of the triceratops.

The long-necked planteater apatosaurus (once called a brontosaurus) lived during the Late Jurassic and Early Cretaceous periods. An apatosaurus could weigh up to 30 tons and stretch as long as 21 m (70 ft). Its great size allowed it to eat leaves from tall trees and helped it fight off meat-eaters.

The triceratops, famous for its three horns and its bony neck frill, lived in the Late Cretaceous. It reached 9 m (30 ft) long and could weigh 5 tons. The triceratops was a plant-eater that could use its beaked mouth to nip food from low branches.

All triceratops dinosaurs lived later than all apatosaurus dinosaurs. No matter where in the world you look, triceratops fossils are in higher rock layers than apatosaurus fossils.

Paleontologists call this the *law of fossil succession. Succession* means "the order of one thing following another." Earth's plants and animals came into existence in the same order all over the world, so their fossils are always found in the same order.

In the Cedar Mountain Formation in Utah, scientists found the teeth of a deinonychus, as well as the remains of a plant-eating dinosaur called the iguanodon. The rock layer that these fossils were found in was from the Early Cretaceous period. Paleontologists have also found deinonychus fossils in Montana and Oklahoma and iguanodon fossils in Europe and Asia. No matter where on Earth deinonychus or iguanodon fossils are found, they're always in the Early Cretaceous rock layers.

Fossil Age = Rock Age

Fossils are used to learn the relative ages of rocks. Tyrannosaurus dinosaurs lived during the Late Cretaceous period. If a layer of sedimentary rock contains a tyrannosaurus fossil, then that rock formed during the Late Cretaceous period. Remember that deinonychus fossils are always found in rock formed during the Early Cretaceous period, so a tyrannosaurus fossil will be newer than the deinonychus fossil. This is true no matter where on Earth these fossils are found.

In fact, the ancient time periods are divided by the kinds of fossils that are found in those time periods. Scientists check fossils to learn the relative age of a rock layer.

Next to the Red Rocks Park is an area called Dinosaur Ridge. Some famous early dinosaur discoveries were made there, including the bones of a long-necked apatosaurus, an armored stegosaurus, and a ferocious allosaurus. These dinosaurs all lived during the Late Jurassic period. This tells geologists that the Red Rocks sandstone dates back to the Late Jurassic period.

You have now learned four ways in which scientists can determine a rock layer's relative age. First, a rock layer is older than layers above it and newer than the layers below. Second, if a layer is tilted, that layer is older than the force that tilted it. Third, if magma has pushed through a layer of rock, then that rock is older than the cooled igneous rock. Finally, the fossils that are within a rock layer can tell scientists that layer's age.

Index Fossils

Although the dinosaurs have died out, some types of early plants and animals have not. Sharks today, for example, are much like the sharks that lived nearly 400 million years ago.

On the other hand, trilobites (TRY•luh•byts) lived almost as long as sharks, but then died out. A *trilobite* was a small animal, usually 3 to 7 cm (1 to 3 in.) long that had a shell covering its back. Many different kinds of trilobites existed, and most of them lived in shallow water or near the shore. Trilobite fossils can be found all over the world.

Scientists are very interested in trilobites because they existed for a known period of time and are found in many places. Scientists consider a trilobite to be an index fossil. *Index fossils* show the age of the rock layer they're in, which tells the relative age for rocks above and below that layer. Any rock layer that has trilobite fossils in it must have formed during the 350 million years that trilobites existed.

Late Cretaceous Period

Early Cretaceous Period

Late Jurassic Period

Another index fossil is the ammonite. Ammonites were sea animals that lived in curled shells. Their size ranged from 2 cm (0.75 in.) to 1.7 m (5.6 ft) across. Ammonites ate small sea organisms and moved by spurting water behind them. They existed between 400 million years ago and 65 million years ago.

Ammonite fossils are found in many places and are easy to recognize. These two traits make them a good index fossils. Paleontologists know that ammonites changed over time. So when an ammonite fossil is found, a paleontologist can determine the age of the rock layer it was found in by how the fossil looks.

Trilobites and ammonites are both index fossils. Trilobites lived 400 million years ago at the bottom of shallow water, and ammonites swam freely.

Ammonite

Trilobite

Index fossils also show the relative age of other fossils. A shark fossil might be found with a trilobite fossil, showing that sharks and trilobites both existed at the same time. A shark fossil could also be found with an ammonite fossil, proof that sharks and ammonites lived at the same time.

In Land Or Sea?

In addition to being an index fossil, the trilobite is important in another way. Like the fossils of other animals that lived in or near water, a trilobite fossil shows that the rock layer holding it was at one time under water.

In the Grand Canyon, the fossils found in the Redwall Limestone, including trilobites, are of sea animals. When that rock layer formed, the area wasn't the desert it is now—it was a shallow sea. The next layer up has fossils of sea animals in one area and land plants in a different section. In the following layer, ferns and trees are found. These fossil layers show that land slowly replaced water.

Radioactive Elements

Using relative dating, scientists can tell the order of rock layers and fossil layers, but how do they know a fossil's real age? How did they learn that sharks were on Earth almost 400 million years ago? These questions can be answered by understanding elements and how they behave.

Earth is made up of different elements. Some examples of elements are oxygen (in the air we breathe), helium (a gas that can make a balloon float), and sodium (a component of salt). Each element is made up of one kind of atom. Oxygen is made up of oxygen atoms, helium is made up of helium atoms, and sodium is made up of sodium atoms.

An element you may not have heard of before is uranium. Uranium is a special element because it is radioactive. *Radioactive* means that its atoms decay. When an atom *decays,* it changes into the atom of a different element. Uranium, for example, decays into the element lead.

Some radioactive elements decay quickly and some decay slowly. The time it takes for half of its atoms to decay is an element's *half-life.*

For example, uranium-235 has a half-life of 713 million years. That means that in 713 million years, half of its atoms will decay into lead atoms.

Radioactive Dating

Uranium and a few other radioactive elements are a normal part of nature. The granite at the base of the Grand Canyon contains some of the mineral zircon. Most zircon contains a small amount of uranium. The longer zircon lies in the ground, the more its uranium atoms decay. Gradually, the zircon has less uranium and more lead.

If a geologist measured the amount of uranium and lead in the zircon, it would tell her how much of the uranium had decayed. The geologist knows the uranium's half-life, so she could count how long the uranium had been decaying.

When geologists study the Grand Canyon, they learn the order of rock layers and how the land has changed. They have done radioactive dating of the granite and schist at the canyon's bottom.

When scientists conduct *radioactive dating* on a rock, they measure a rock for how much of its uranium or other radioactive elements have decayed. This tells the age of the rock.

Uranium and other radioactive elements don't decay rapidly enough to date relatively recent events, such as the formation of the volcano Paricutín in 1943. But radioactive dating is useful for dating rocks that are millions of years old.

Dating Igneous Rocks

Within magma, uranium separates from the lead that it had decayed into. When the magma cools, the uranium's slow decay takes place in the new igneous rock. Millions of years later, when geologists measure the amounts of uranium and lead, they can learn the date the magma became solid.

Geologists conduct radioactive dating on rocks that have been underground. If a rock lies on Earth's surface for a million years, some of its uranium or lead will have weathered away.

Scientists have used radioactive dating to learn the ages of rock formations all over the world. They measured a granite rock in the White Mountains of New Hampshire as being 180 million years old. A mineral sample found on Mount Rogers, Virginia, was 820 million years old. An ancient metamorphic rock in Minnesota turned out to be 3.6 billion years old.

Radioactive dating has shown that Earth probably came into being more than 4.5 billion years ago. This date didn't come from any Earth rock. Scientists measured both a meteorite and a moon rock as being 4.5 billion years old. Because the moon came into being at the same time as Earth, its rocks are about the same age as Earth.

Dating Sedimentary Layers

Learning the age of sedimentary rocks is more complicated. Sediment is a blend of igneous, sedimentary, and metamorphic rocks weathered into small pieces. These pieces mix older and newer rock into a layer. Scientists can use radioactive dating to learn the age of the older and newer rock, but not to find out when the sediment as a layer turned to rock.

Geologists look for places where sedimentary rock lies just above or below igneous rock. The igneous rock might be an ancient lava flow or a layer of volcanic ash. Geologists use radioactive dating to calculate the age of the igneous rock, knowing that a sedimentary layer is older than igneous rock above it but newer than the igneous rock below.

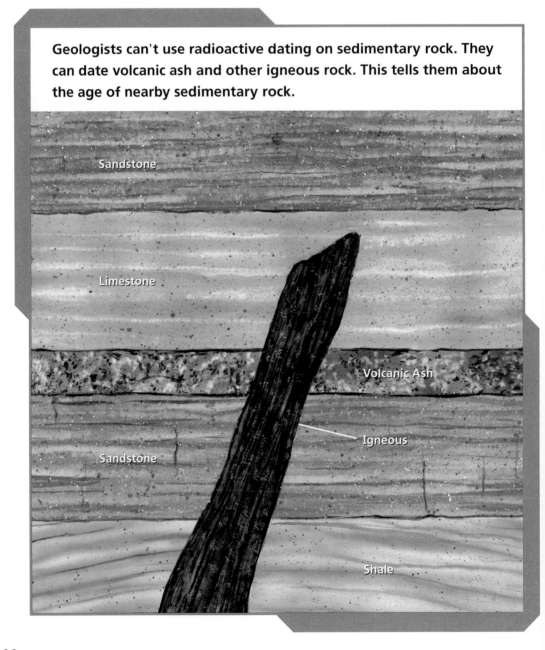

Geologists can't use radioactive dating on sedimentary rock. They can date volcanic ash and other igneous rock. This tells them about the age of nearby sedimentary rock.

Sandstone

Limestone

Volcanic Ash

Igneous

Sandstone

Shale

Geologists look for igneous rock that, when it was hot magma, pushed through several layers of sedimentary rock. They know that all of the sedimentary layers are older than the magma that moved through them. When they conduct radioactive dating of that igneous rock, they learn more about the age of the sedimentary rock layers.

At the Grand Canyon, the bottom layer of metamorphic schist was cut by an ancient block of granite, which is igneous rock. That granite has been dated as being 1,500 million years old. From this, geologists know that the schist must be older than that. In fact, radioactive dating put its last metamorphic change at 1,675 million years ago.

Above the granite and schist is a series of tilted sedimentary layers called the Unkar Group. This layer is made of gray limestone with some shale. The lowest layer of it has to be newer than the schist, so it is less than 1,675 million years old. Geologists have determined its relative age to be about 1,250 million years.

When Did Deinonychus Live?

Using radioactive dating, geologists have calculated the ages of thousands of rock layers around the world. In doing so, they have found the ages of nearby sedimentary layers and of the fossils inside those layers. Best of all, they have found the age of the index fossils.

Deinonychus dinosaurs lived during the Early Cretaceous period. Paleontologists in Montana found five complete deinonychus skeletons next to a large plant-eating dinosaur. No one knows whether the bodies were just caught in the same flood, or if a pack of deinonychus died while attacking the plant-eater. Scientists do know that these animals lived between 144 and 98 million years ago.

The deinonychus lived between
144 and 98 million years ago.